AF599086

★ MILITARY VEHICLES ★

# DRONES

MARTY GITLIN

2001 SW 31st Avenue
Hallandale, FL 33009
www.mitchelllane.com

First Edition, 2021.

Author: Marty Gitlin
Designer: Ed Morgan
Editor: Morgan Brody

Little Mitchie is an imprint of Mitchell Lane Publishers.

Title: Military Vehicles: Drones / by Marty Gitlin
Description: Hallandale, FL :
Mitchell Lane Publishers, [2021]

Series: Military Vehicles
Library bound ISBN: 978-1-58415-290-3
eBook ISBN: 978-1-58415-298-9

Photo credits: Freepik.com, Freevector.com, p. 17 U.S. Air Force photo/Tech. Sgt. Christopher Boitz, p. 18 Ssgt. Vernon Young Jr/ZUMA Press/Newscom, p. 19 Photo by David Henrichs on Unsplash

# CONTENTS



Words in **bold** can be found in the Glossary.

# 1

# THE BALLOONS OF WAR

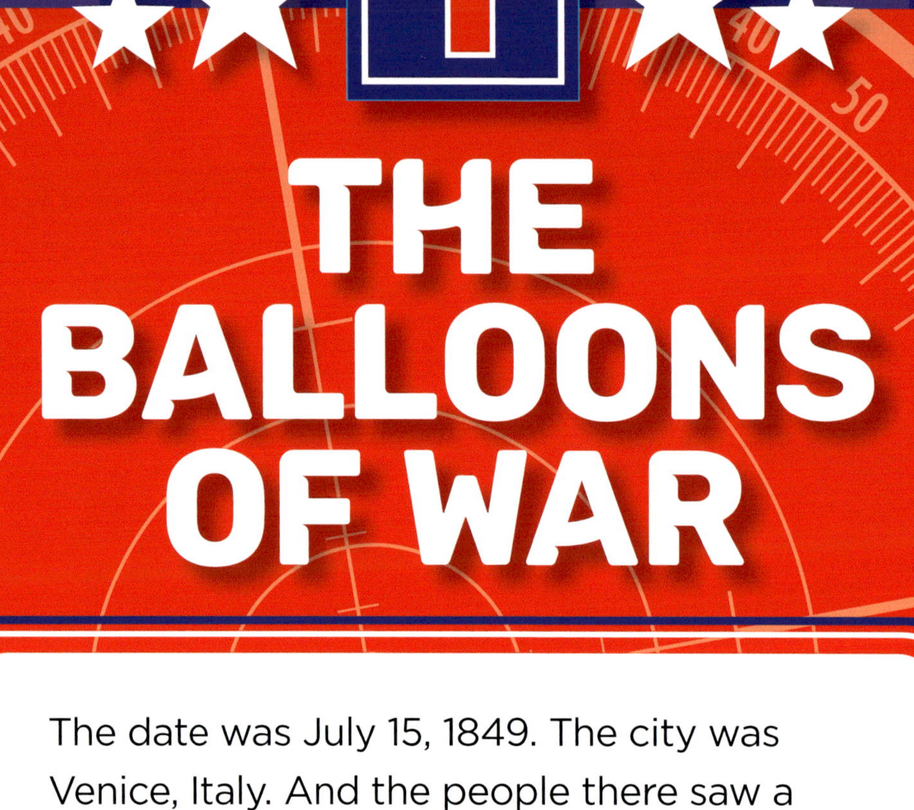

The date was July 15, 1849. The city was Venice, Italy. And the people there saw a strange sight. They watched huge hot-air balloons float in the sky.

Those craft came courtesy of Austria. That nation was at war with Italy. The **unmanned** balloons released bombs filled with **shrapnel**. They did little damage to Venice.

The balloons were drones. A drone is any air vehicle without a pilot. There were no pilots in 1849. There were no airplanes.

Winged craft later changed warfare forever. Most of them were manned. Their pilots dropped bombs and battled enemy planes in the sky.

## FAST FACT

### SKETCHY DETAILS

Details vary about the balloon attack on Venice. One account claims about 200 balloons filled the sky. Another states there were only two. It is believed the bombs were released or blown up by a timer.

# 2

# PLANES WITHOUT A PILOT

It was 1916. World War I raged in Europe. The **British** wanted control of the skies. That country developed the first unmanned winged airplane. It was called the **Aerial** Target. And it was a drone.

The **craft** was a flying bomb. It was intended to explode at enemy aircraft. It was also made to hit ground targets. But testing failed. The British gave up. They believed drones had little military value.

They were wrong. American Charles Kettering proved it. He invented his own unmanned plane. It was called the Kettering Bug.

The Kettering Bug.

The Bug had controls that guided it toward targets. One control shut off the engine. The wings then released. The craft would **plunge** to earth and explode at the target.

The Kettering Bug proved that drones could have a military impact.

Kettering Aerial Torpedo, nicknamed the "Bug."

## FAST FACT

# A BIT ABOUT THE BUG

The Dayton-Wright Airplane Company built many Kettering Bugs during World War I. But it was too late to use them in combat. The war ended in 1918.

# 3

# WHAT DRONES DO

The role of unmanned aircraft has **evolved**. Drones have gone from nearly useless to **vital**.

The U.S. military has made that happen. It used radio-controlled craft for target practice in the 1930s. Drones with TV receivers took photos of bomber targets during World War II. Others were developed to drop bombs.

Those drones had limited success. So did those used by the U.S. and Russia to spy on each other in the 1950s.

American drones soon played a role in the Vietnam War. They acted as **decoys** in combat. They launched **missiles** on targets.

The U.S. Air Force Ryan AQM-34L *Firebee* drone "Tom Cat."

The modern era of drones began in the 1980s. They have taken lives and saved lives. They **monitor** climate change. They search for people trapped after earthquakes. They deliver packages.

Military drones get the most attention. They find targets for strikes. They **survey** areas where troops cannot go safely. They fire missiles at the enemy.

A BQM-74E aerial drone is launched from the guided-missile frigate USS *Thach*.

The General Atomics MQ-1C *Sky Warrior.*

## FAST FACT

### NOT ALWAYS PERFECT

The U.S. military tries to use drones safely. But drone strikes do not always end well. Some have hit innocent people. The military has been criticized for such mistakes.

# 4

# HOW DO THEY WORK

Drones can be 20 pounds. Or they can be 1,320 pounds. They can fly under 1,200 feet in the air. Or they can fly 18,000 feet high. They can travel less than 100 miles an hour. But some can fly an amazing 17,000 miles an hour.

They have names such as the *Predator, Reaper, Global Hawk*, and *Black Hornet Nano*. Some are small enough to take off from a **launcher**. Others are flung from an air strip.

An airman escorts an RQ-4 Global Hawk back to a hangar.

Some military drones hold camera **sensors** to survey. They take pictures to find the enemy. Some stay in the air for a few minutes. Other fly in the sky for up to 32 hours.

The U.S. military has worked to find more uses. They are the wave of the future.

Drone operators look at information sent from the drones.

## FAST FACT

### DRONES ARE TAKING OFF

Drones have become popular with people outside the military. There were about two million drones in the U.S. in 2019. Many people use them as a hobby.

# 5

# EYE TO THE FUTURE

One wonders what the future of drones will look like.

Some believe the military will build larger drones. Others feel smaller and faster drones are in the future. But everyone hopes they are used more for peace than war.

Drones have been used more often in recent years to spy. Some innocent people have been killed by drone strikes. A U.S. drone was shot down in 2019 by the Iranian military. That caused a problem between the two countries.

How will nations deal with drones? Will they be used to sew mistrust? Or will they be used to help people? Only time will tell.

## FAST FACT

### A DRONE TAXI?

A company in China hopes to turn drones into cabs. It built and tested an unmanned craft. The drone would serve passengers as do Uber and Lyft. *But they would fly!* And they would get people around much faster.

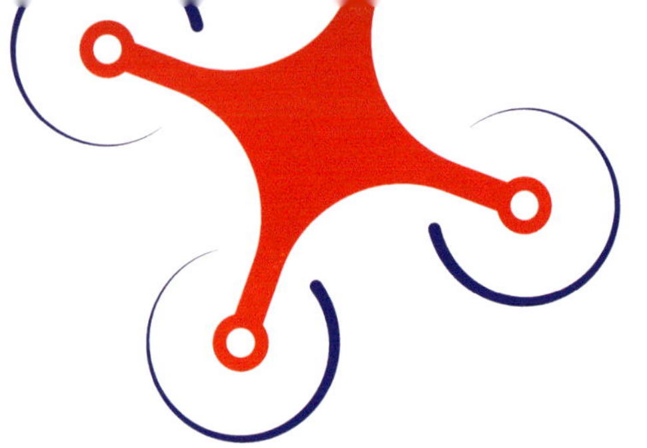

# What You Should Know

- Drones come in many shapes and sizes.
- Many military drones are not used as weapons.
- Some drones can fit in the palm of a hand while others weigh over 1,000 pounds.
- The definition of a drone is an aerial craft without a pilot.
- Drones can be controlled by radio signals.
- The first military drones were balloons.
- The modern military drone was not launched until 1982.
- The fastest drone can travel nearly 17,000 miles an hour.

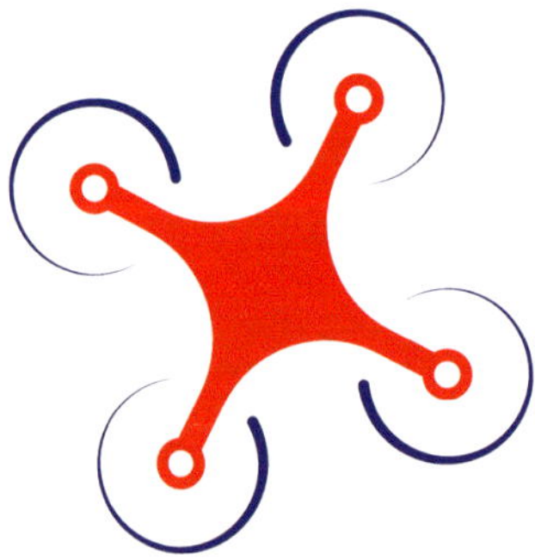

# Glossary

**aerial**
Related to aircraft

**British**
People from England

**craft**
An airplane, helicopter, or spacecraft

**decoy**
Something intended to lure into a trap

**evolve**
To develop or grow

**launcher**
To send an object with force

**missile**
Object shot or launched to strike and damage a target, often by exploding

**monitor**
Watch or observe in keeping track of something

**plunge**
To fall or drop suddenly

**sensor**
Device that detects something and responds by transmitting a signal

**shrapnel**
Metal pieces from an exploded shell

**survey**
To look over and examine closely

**unmanned**
Without a pilot

**vital**
Extremely important

# Find Out More

Marsico, Katie. *Drones (A True Book: Engineering Wonders)*. Chicago, IL: Children's Press, 2016.

Baby Professor. *How Do Drones Work?* Technology Book for Kids.

Mairghread, Scott. *Science Comics: Robots and Drones: Past, Present, and Future*. New York, N.Y.: Macmillan Publishers, 2018.

# Websites

**Ducksters: This page on the education site teaches kids about the U.S. military.**
https://www.ducksters.com/history/us_government/united_states_armed_forces.php

**Britannica Kids: Learn all about drones in the video on this site.**
https://kids.britannica.com/kids/assembly/view/214255

**Kiddie Encyclopedia: The page here provides facts for kids about the U.S. Air Force.**
https://kids.kiddle.co/United_States_Air_Force

# Index



# About the Author

Martin Gitlin is a freelance author living in Cleveland, Ohio. He has written about 150 books since 2006, mostly for young students. He embraces military history and boasts a strong interest in drones. Martin worked as a newspaper writer from 1991 to 2002. He won more than 45 awards during that time. Included was a first place for general excellence from Associated Press in 1996. That group also voted him as one of the top four feature writers in Ohio in 2002.